MW01102156

Exploring Planets
PLUTO

Susan Ring

WEIGL PUBLISHERS INC.

Published by Weigl Publishers Inc.
350 5th Avenue, Suite 3304, PMB 6G
New York, NY USA 10118-0069
Web site: www.weigl.com

Library of Congress Cataloging-in-Publication Data

Ring, Susan.
Pluto / by Susan Ring.
v. cm. -- (Exploring planets)
Includes index.
Contents: Introducing Pluto -- What's in a name? -- Pluto spotting -- Early observations -- Pluto in the solar system -- Pluto and Earth -- Missions to Pluto -- Pluto explorer: Clyde Tombaugh -- Pluto explorer: James Christy -- Pluto on the web -- Activity: Pluto math -- What have you learned?
ISBN 1-59036-101-6 (lib. bdg. : alk. paper) – ISBN 1-59036-228-4 (pbk.)
1. Pluto (Planet)--Juvenile literature. [1. Pluto (Planet)] I. Title. II. Series.
QB701 .R56 2003
523.482--dc21

2002014499

Printed in the United States of America
1 2 3 4 5 6 7 8 9 0 08 07 06 05 04

Photograph Credits

Every reasonable effort has been made to trace ownership and to obtain permission to reprint copyright material. The publishers would be pleased to have any errors or omissions brought to their attention so that they may be corrected in subsequent printings.

Cover: The Image Bank by Getty Images (top); Digital Vision (bottom)

Virginia Boulay: pages 8, 12; **CORBIS/MAGMA:** pages 11 (Dave G. Houser), 18 (Bettmann); **Hulton|Archive by Getty Images:** page 6; **The Image Bank by Getty Images:** pages 1, 3, 4, 22; **Johns Hopkins University Applied Physics Laboratory/Southwest Research Institute:** page 17; **NASA:** pages 10, 13, 14, 16; **Tom Stack & Associates:** pages 7 (TSADO/M. Buie-Lowell Observatory), 9 (TSADO/NASA), 19 (NASA/ESA).

Project Coordinator Jennifer Nault **Design** Terry Paulhus
Substantive Editor Tina Schwartzenberger **Copy Editor** Heather Kissock
Layout Bryan Pezzi **Photo Researcher** Tina Schwartzenberger

Contents

Introducing Pluto

Pluto is the farthest planet from the Sun. It was the last of the nine planets to be discovered. Due to its distance from Earth, it has never been visited by a **space probe**. Today, scientists are just beginning to find answers to their questions about Pluto. Read on to learn about the planet that was discovered by accident.

The closest picture of Pluto was taken when it was 2.6 billion miles (4.4 billion kilometers) from Earth.

Pluto Facts

- Pluto moves slower in its **orbit** than any other planet.
- Pluto is a tiny planet. It is smaller than seven of the moons in our **solar system**.
- Some areas on Pluto are very dark, while other areas are very bright.
- The ice on Pluto's surface is not made of water. It is made of **nitrogen**.
- Pluto only has an **atmosphere** when it moves close to the Sun. The heat of the Sun melts Pluto's icecaps. The icecaps become gas.
- Some scientists believe that Pluto is not a planet at all. It is not made of rock like the four planets closest to the Sun. It is not made of gas like the remaining planets. Instead, Pluto is made of rock and ice.

Name That Planet

Pluto is named for a god in Greek **mythology**. This god ruled the **underworld**.

When the planet was discovered in 1930, many people came up with possible names for it. Some of the names suggested included Atlas, Vulcan, and Zeus. Finally, an 11-year old girl came up with the name Pluto. The first two letters of the word *Pluto* are the initials of the **astronomer** who helped discover the planet. His name was Percival Lowell.

Some people think Pluto was named after the god of the underworld because the planet is far away from the Sun and is always in darkness.

Pluto Moons

Pluto's only moon is called Charon. It was discovered in 1978. Pluto and its moon are similar in size. Their orbits are also very close. This is why Pluto has been called the "double planet."

Charon is covered in ice. Astronomers think that Charon was once a part of Pluto.

The *Hubble Space Telescope's* photographs of Pluto and Charon show that the moon reflects less light than the planet does.

Pluto Spotting

Pluto loses its place as the ninth planet from the Sun every 248 years. At this time, it becomes the eighth planet from the Sun. This happens because Pluto's orbit is unlike the orbits of the other planets in our solar system. The other planets move in a circular orbit. Pluto's orbit is not a circle. Instead, it moves in an **elliptical orbit**. This type of orbit is oval-shaped. Pluto's different orbit causes it to cut in front of Neptune for about 20 years.

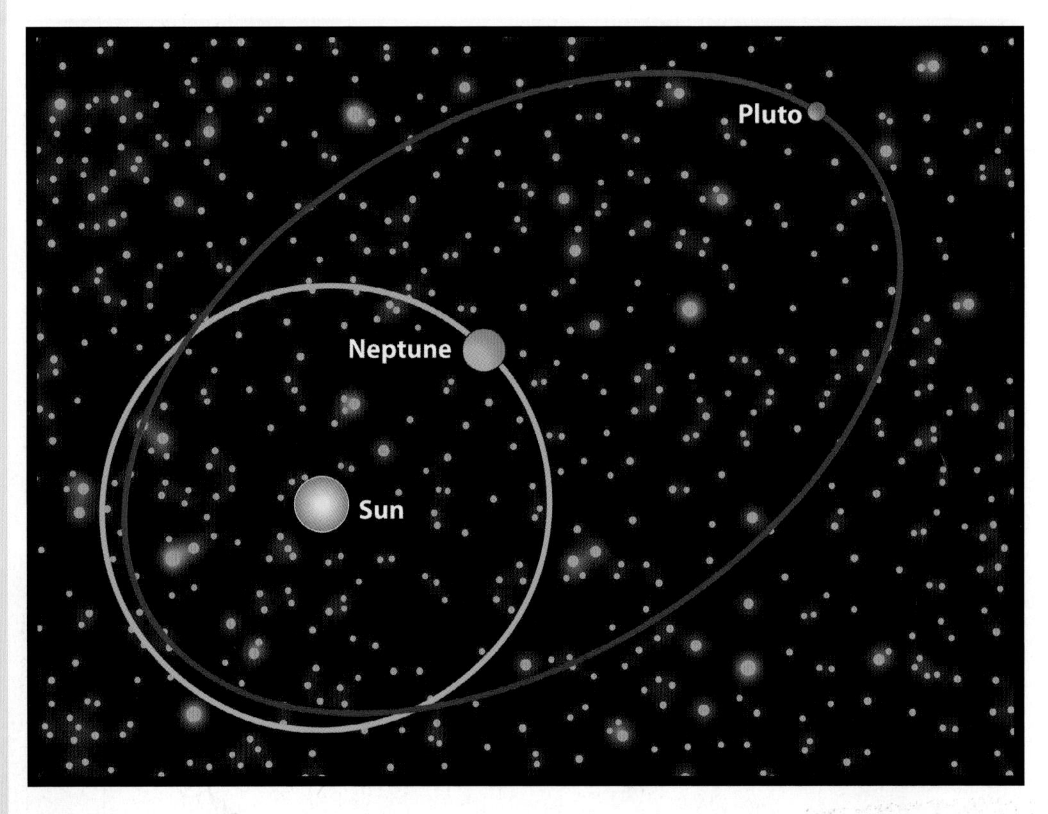

Pluto was last inside Neptune's orbit from 1979 through 1999.

See for Yourself

Many small, frozen objects orbit the Sun beyond Neptune and Pluto. These bodies are called trans-Neptunians. Some astronomers think that these objects could be comets. This is why some people believe that Pluto is not really a planet. They think that it may be a comet, like the other space objects nearby.

You cannot see Pluto with just your eyes. The planet is too far away from Earth. Pluto can sometimes be seen through a telescope. It looks like a speck in the sky. Ask science center staff for the best times to view Pluto.

Even the clearest photographs of Pluto do not show details of the planet.

Early Observations

For many years, astronomers thought that another planet existed beyond Uranus. If so, this would explain why Uranus's orbit was so unusual. The **gravity** of another planet would affect Uranus's orbit. Astronomers proved this **theory** correct when they found Neptune.

Astronomers began looking for another planet past Neptune. Mathematics suggested that Neptune's orbit was affected by the gravity of another planet. The math turned out to be wrong, but Pluto was found anyway.

Neptune's mass is over 8,000 times greater than Pluto's mass.

Planet X

Astronomers began to search for a planet beyond Neptune. This mysterious planet was called Planet X. Astronomer Percival Lowell began looking for Planet X. He even built his own **observatory** in Arizona to study outer space. Percival spotted Planet X through his telescope. At first, he did not realize that he had found Planet X.

Pluto was first observed from the Pluto Dome at the Lowell Observatory.

Pluto in Our Solar System

Pluto is one of the nine planets in our solar system. It is the farthest planet from the Sun.

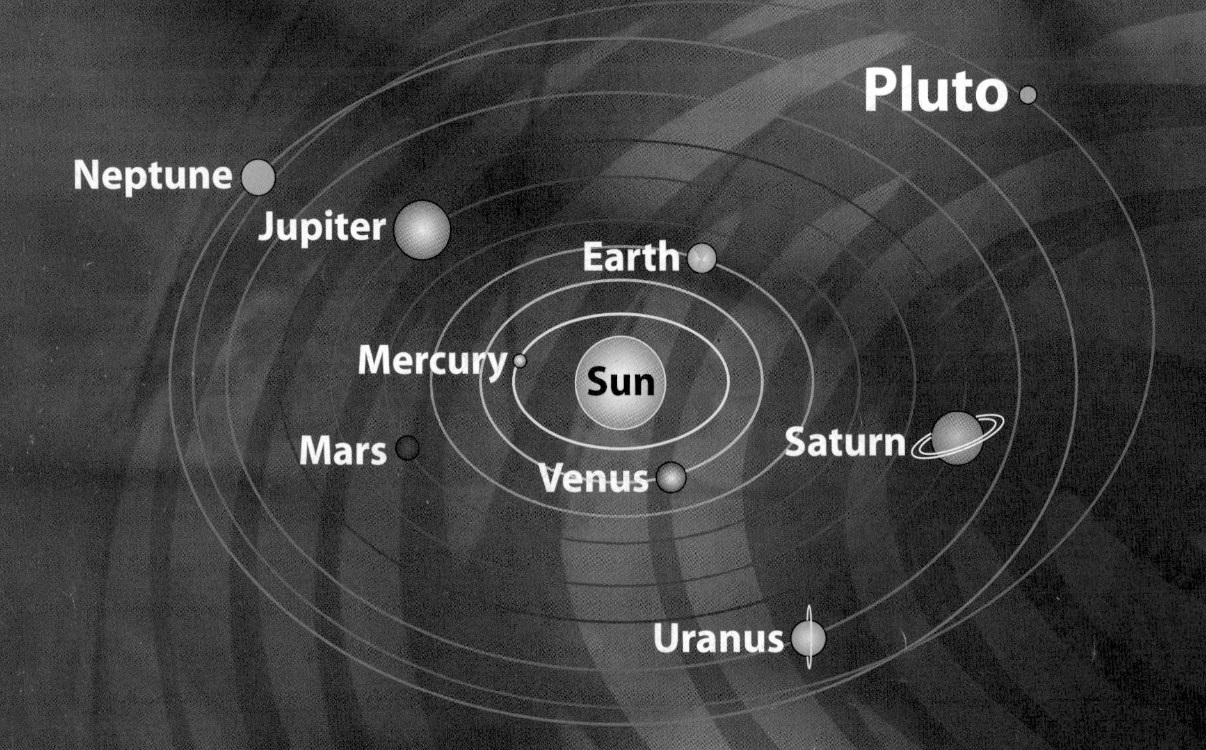

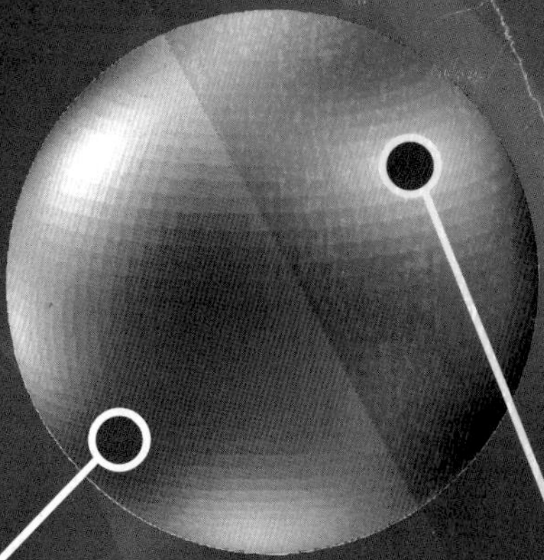

There are light spots on Pluto. These spots are likely nitrogen ice.

The darker areas of Pluto are thought to have a warmer temperature than the lighter areas.

Pluto and Earth

Pluto is much smaller than Earth. In fact, Earth is about five times larger than this distant planet. Even our Moon is larger than Pluto.

Not much is known about Pluto's atmosphere. Astronomers think that it is mainly made of nitrogen. For much of Pluto's year, the atmosphere is frozen ice.

Astronomers have discovered that Pluto's atmosphere contains small amounts of carbon dioxide and methane gases.

Compare the Planets

PLANET FEATURES					
PLANET	Distance from the Sun	Days to Orbit the Sun	Diameter	Length of Day	Average Temperature
Mercury	36 million miles (58 million km)	88	3,032 miles (4,880 km)	4,223 hours	333° Fahrenheit (167° C)
Venus	67 million miles (108 million km)	225	7,521 miles (12,104 km)	2,802 hours	867° Fahrenheit (464° C)
Earth	93 million miles (150 million km)	365	7,926 miles (12,756 km)	24 hours	59° Fahrenheit (15° C)
Mars	142 million miles (229 million km)	687	4,222 miles (6,975 km)	25 hours	–81° Fahrenheit (–63° C)
Jupiter	484 million miles (779 million km)	4,331	88,846 miles (142,984 km)	10 hours	–230° Fahrenheit (–146° C)
Saturn	891 million miles (1,434 million km)	10,747	74,897 miles (120,535 km)	11 hours	–285° Fahrenheit (–176° C)
Uranus	1,785 million miles (2,873 million km)	30,589	31,763 miles (51,118 km)	17 hours	–355° Fahrenheit (–215° C)
Neptune	2,793 million miles (4,495 million km)	59,800	30,775 miles (49,528 km)	16 hours	–355° Fahrenheit (–215° C)
Pluto	3,647 million miles (5,869 million km)	90,588	1,485 miles (2,390 km)	153 hours	–375° Fahrenheit (–226° C)

Missions to Pluto

There have not been any space missions near Pluto. Pluto's great distance from Earth makes getting there difficult. The *Hubble Space Telescope* has taken the best photographs of the planet. Still, these images are blurry. They only show the largest features of the planet.

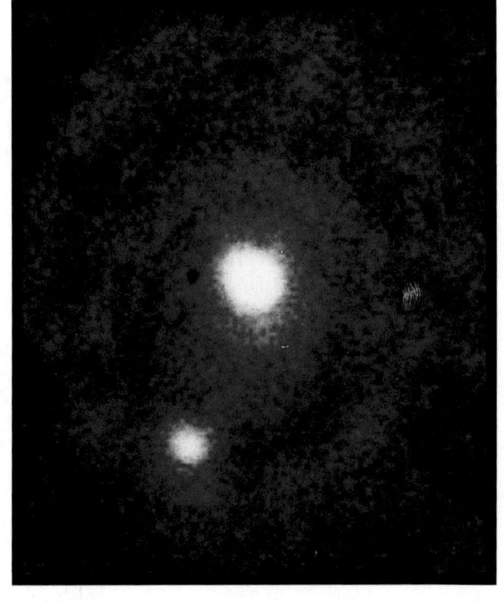

■ The photograph on the left was taken by a ground-based telescope. The one on the right was taken by the *Hubble Space Telescope.* Although the Hubble photograph is clearer, it still does not show Pluto's or Charon's surface.

Pluto Plans

There is a space mission to Pluto planned for the near future. This mission is called New Horizons. A spacecraft will fly by Pluto and Charon after about 9 years of travel. No one will be aboard the spacecraft. It will be controlled from Earth. The New Horizons mission will take thousands of photographs. These pictures may provide more information about Pluto's surface, atmosphere, and weather.

New Horizons should arrive at Pluto and Charon in 2015.

Planet People

Pluto Explorer: Clyde Tombaugh

Name: Clyde Tombaugh
Pluto Accomplishments: Discovered Planet X, known today as Pluto

American Clyde Tombaugh found Planet X in 1930. He was only 24 years old at the time of the discovery. As a child, Clyde shared his father's interest in **astronomy**. After high school, Clyde did not have enough money to attend college. Instead, he worked at the Lowell Observatory. He used the giant telescope there to photograph millions of stars. One day, Clyde noticed something unusual in one of the photographs. An object thought to be a star had moved. While stars do not appear to move in space, planets are always moving around the Sun. Clyde had discovered Planet X. Soon after the discovery, Planet X was named Pluto.

Clyde Tombaugh was an assistant at the Lowell Observatory when he discovered Pluto.

Pluto Explorer: James Christy

Name: James Christy
Pluto Accomplishments: Discovered Charon, Pluto's moon

In 1978, James Christy discovered that Pluto has a moon. While looking at the planet through his telescope, he saw an object that looked like a bump. He decided to watch this bump for a while. Soon, he realized that the bump moved. When the bump moved, James knew that he had spotted a moon. James named the moon Charon.

The *Hubble Space Telescope* has provided a much clearer image of Pluto (top) and Charon than the "bump" James Christy saw with his telescope.

Pluto on the Internet

To learn more about Pluto, look for books at your school library. The Internet is also an excellent place to learn about Pluto. There are many great Web sites with information. Just type the words *Pluto* and *planet* into a search engine. Google and Yahoo are useful search engines.

The Internet has information on all of the planets in our solar system. To learn about the nine planets, visit these Web sites:

Encarta Homepage
www.encarta.com
Type the name of a planet that you would like to learn about into the search engine.

NASA Kids
http://kids.msfc.nasa.gov
NASA built a Web site for young learners just like you. Visit this site to learn more about the nine planets, space travel, and the latest NASA news.

Young Scientists at Work

Try this activity to see how far Pluto is from the Sun compared to Earth.

You will need:

- a ball, such as a beach ball
- string
- scissors and tape

In this activity, the ball stands for the Sun. Pieces of string stand for the distance between the planets and the Sun.

Measure a piece of string along your arm, from your shoulder to your fingertips. Cut the string to be this length. Place the large ball on the floor. Tape the string to the bottom of the ball, and stretch it out along the floor. This piece of string stands for the distance from the Sun to Earth.

Next, cut a piece of string forty times longer than the first piece. Tape it to the end of the first piece of string. Stretch it out along the floor. Look at the difference of the length of the strings. This represents how far Pluto is from the Sun compared to Earth.

What Have You Learned?

How much do you know about Pluto?
Test your knowledge!

1 **True or False? Pluto is the third-largest planet in our solar system.**

2 **What is the name of the space mission heading to Pluto in the near future?**

3 **How many space probes have been to Pluto?**

4 **When does Pluto have an atmosphere?**

5 **What is the name of the observatory where Pluto was discovered?**

6 **Who discovered Pluto?**

7 What was the first name given to Pluto?

8 What is the name of Pluto's moon?

9 Why is Pluto sometimes called the double planet?

10 What happens to Pluto's orbit every 248 years?

What was your score?

9–10	You should work at NASA!
5–8	Not too bad for an earthling!
0–4	You need to polish your telescope!

Answers

1 False. Pluto is the smallest planet. **2** The space mission is called New Horizons. **3** No space probes have ever been to Pluto. **4** Pluto has an atmosphere when it gets close to the Sun. **5** Pluto was discovered at Lowell Observatory. **6** Clyde Tombaugh discovered Pluto. **7** Pluto was first called Planet X. **8** Pluto's moon is called Charon. **9** Pluto is sometimes called the double planet because it orbits so close to its moon. **10** Every 248 years, Pluto crosses Neptune's path and becomes the eighth planet from the Sun.

Words to Know

astronomer: a person who studies space and its objects

astronomy: the study of planets, stars, and other objects in space

atmosphere: the layer of gases surrounding a planet

elliptical orbit: an orbit in the shape of an oval

gravity: a force that pulls things toward the center

mythology: stories or legends, often about gods or heroes

nitrogen: a gas without color or odor

observatory: a building with a large telescope

orbit: the nearly circular path a space object makes around another object in space

solar system: the Sun, the planets, and other objects that move around the Sun

space probe: a spacecraft used to gather information about space

theory: a guess based on knowledge

underworld: in Greek mythology, the place where the souls of the dead go

Index